James S. Tate

Surcharged and different forms of retaining walls

.

James S. Tate

Surcharged and different forms of retaining walls

ISBN/EAN: 9783337717957

Printed in Europe, USA, Canada, Australia, Japan

Cover: Foto ©ninafisch / pixelio.de

More available books at **www.hansebooks.com**

SURCHARGED

AND

DIFFERENT FORMS

OF

RETAINING WALLS.

BY

JAMES S. TATE, C. E.

NEW YORK:

D. VAN NOSTRAND, PUBLISHER,

23 MURRAY AND 27 WARREN STREET.

1873.

PREFACE.

This little Work is intended to supply what has no doubt been often wanted by many Engineers—a certain and ready means of correctly and easily ascertaining the Pressures of Embankments, Submerged or otherwise, composed of different materials; also the Moments of Retaining Walls of different forms of cross-section, to successfully withstand those pressures; so that, by knowing the exact value of each, the right dimensions of the most suitable form of wall for the purpose required can be at once ascertained.

As the method adopted does not involve the use of any long or laborious calculations, it is hoped it will prove useful to the Profession generally.

RETAINING WALLS.

Retaining walls are adopted as a necessary expedient in railway and other practice, often under very peculiar circumstances, as when there is not sufficient room for the slope of the embankment; it being sometimes perched high on a steep mountain's side, and where it would have been hardly possible to construct a railway at all, except by securing it with a massive wall occupying comparatively little space.

When it is also remembered how fearfully terrible any accident would be if it was to occur in such a dangerous situation—if by any erroneous calculation or mistaken judgment on the part of the engineer sufficient strength had not been given to the work, the wall which was to have supported the embankment, suddenly giving way, falling over into a deep ravine or chasm, a large portion of the embankment going with it, and, it may be also, a passing train —there can be no doubt but that the nature of the material of which the embankment is to be made should be understood, and the best form and requisite dimensions for the wall should be well considered and accurately ascertained beforehand, so that it may be amply strong enough.

At the same time that the wall should be made perfectly secure, it is also often desirable that any unnecessary excess of strength should not be given to it, and so thereby avoid increasing its cost considerably, as the value of work is often very much enhanced when it has to be executed in such inaccessible situations as before mentioned, where all the materials for building it may have to be brought from a great distance.

The engineer thus may be at a loss to determine of what size a retaining wall should be built, so as to be safe against all contingencies that can occur, and yet also to be economical.

In many cases there have been failures which may have arisen from not correctly ascertaining beforehand how the material of which the embankment is composed will be affected by the alternations of wet and dry weather before it is thoroughly consolidated, and the precise angle at which its slope will stand in either case, thereby causing a considerable difference in its pressure against the wall.

A retaining wall also, as in the case of the wing-walls of a bridge, being built at the same time that the embankment is being filled in behind it, has often to withstand then a considerable greater pressure than it will have to do afterwards when the embankment is settled; this also perhaps when its work is green, and not prepared to resist the pressure intended for it. Sometimes also the punning of the material behind it has (as is often the case) not been done effectually, and a heavy rain changes the dry earth or clay into a wet sludge, causing it to swell considerably.

It therefore being such an important point in railway construction, it would no doubt be very desirable if some simple form of calculation were used, not only strictly accurate, but easily adapted to any circumstances that may occur. In the case of a wall where the embankment is level with its top the calculation of the pressure is well known, being very simple, and is as follows:

Let B D be the back of a retaining wall, D E the natural slope of the embankment,

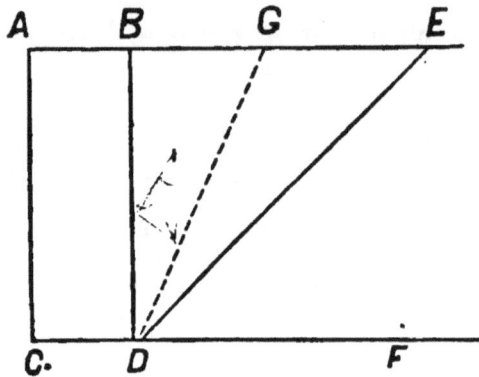

then if we bisect the $\angle$ B D E by the line D G, B D G is the portion of the embankment supported by the retaining wall.

Now the weight of B D G : pressure of its weight against the wall : : B D :.B G : : H : H tang. $\angle$ B D G. The weight of

$$BDG = \frac{H}{2} \times BG \times W =$$

$$\frac{H}{2} \times H \text{ tang. } \angle BDG \times W =$$

$$H^2 \times \frac{\text{tang. } \angle BDG}{2} \times W.$$

Pressure of weight of

$$BDG = H^2 \times \frac{\text{tang. } \angle BDG}{2} \times W \times \text{tang. } \angle BDG$$

$$= H^2 \times \frac{\text{tang.}^2 \angle BDG}{2} \times W.$$

Moment of pressure of weight of

$$BDG = H^2 \times \frac{\text{tang.}^2 \angle BDG}{2} \times W \times \frac{H}{3}$$

$$= \frac{H^3}{6} \times \text{tang.}^2 \angle BDG \times W,$$

and the double of this moment for stability

$$= \frac{H^3}{3} \times \text{tang.}^2 \angle BDG \times W.$$

In the case of a vertical wall, as A B C D, its weight $= W H B$, and the moment of its weight

$$= \frac{W H B^2}{2};$$

then for equilibrium,

$$\frac{W H B^2}{2} = \frac{H^3}{6} \times \text{tang.}^2 \angle BDG \times W,$$

$$\text{and } B = H \text{ tang. } \angle BDG \frac{\sqrt{\frac{W}{3}}}{\sqrt{W}},$$

and for stability,

$$\frac{W\,H\,B^2}{2} = \frac{H^3}{3} \times tang.^2 \angle\,B\,D\,G \times W,$$

$$\text{and } B = H\ tang.\ \angle\,B\,D\,G\ \frac{\sqrt{\dfrac{2\,W}{3}}}{\sqrt{W}}.$$

The figures in the columns of Table No. 1, are calculated from this last formula, and are

$$H\ tang.\ \angle\,B\,D\,G\ \sqrt{\frac{2\,W}{3}},$$

so if divided by the square root of the weight of a cubic foot of the wall, they will give the thickness of the wall.

Table No. 2 gives double the moments of the pressure of the weight of different materials to form the embankment, calculated from the formula

$$\frac{H^3}{3} \times tang.^2 \angle\,B\,D\,G \times W,$$

and which, if made equal to either of the moments of the weight of different forms of retaining walls given, the dimensions of that form of retaining wall required can be readily ascertained.

Having now given the usual formulæ and

Tables for easy calculation deduced from them, for calculating the dimensions of a retaining wall with an embankment level with its top, what is next required is a convenient and ready method of accurately calculating the pressure of a surcharged embankment. The author is not aware if the method of calculation and formulæ he gives here are new, but the Tables for general use have, he thinks, the merit of simplicity.

When the embankment slopes away upwards above the top of the wall, the calculation of its pressure is a little more complex, and no method of finding it has yet been given that is simple, or that can be easily used in practice. Moseley, Hann, and Rankine, in their works give equations very abstruse, and apparently of no practical application. Hann also takes into account the pressure of the slope of the embankment resting on the top of the wall, a refinement of the calculation practically altogether unnecessary, and which, by complicating the original equation, renders mistakes more likely to occur.

If A C be the natural slope of the embankment rising upwards above the top of the wall A B G H, B E a line parallel to it from the foot of the wall, B C bisecting the ∠ A B E, then A B C is the portion of the embankment to be retained by the wall. Now when A B is vertical, the length of the slope to be retained, A C, will be equal to the height of the wall. If ∠ E B F = the angle of the slope of the embankment = ϕ, then

$$\angle \text{A B C} - \angle \text{A C B} = \frac{90° - \phi}{2},$$

and if H = height of the wall, then

$$\text{B C} = \text{H} \frac{\sin. (90° + \phi)}{\sin \left(\frac{90° - \phi}{2} \right)} = \text{H}\,\theta; \quad \text{A L} = \text{H}\sqrt{1 - \frac{\theta^2}{4}},$$

and the weight of

$$\text{B A C} = \frac{W\text{H}^2}{2}\,\theta\,\sqrt{1 - \frac{\theta^2}{4}},$$

W being the weight of a cubic foot of the embankment. Pressure of the weight of

$$\text{B A C} = \frac{W\text{H}^2}{2}\,\theta\,\sqrt{1 - \frac{\theta^2}{4}} \times \text{tang.}\,\frac{90° - \phi}{2},$$

moment of pressure of weight of

$$BAC = \frac{W H^2}{2} \theta \sqrt{1 - \frac{\theta^2}{4}} \times \mathrm{tang.} \frac{90° - \phi}{2} \times \frac{H}{3} =$$

$$\frac{W H^3}{6} \theta \sqrt{1 - \frac{\theta^2}{4}} \times \mathrm{tang.} \frac{90° - \phi}{2},$$

double this moment for stability

$$= \frac{W H^3}{3} \theta \sqrt{1 - \frac{\theta^2}{4}} \times \mathrm{tang.} \frac{90° - \phi}{2}.$$

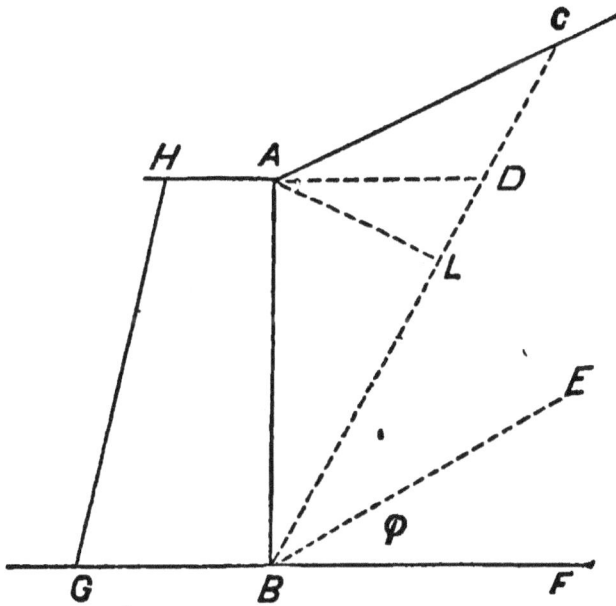

Table No. 3 gives the value of

$$\left(\theta \sqrt{1 - \frac{\theta^2}{4}} \times \mathrm{tang.} \frac{90° - \phi}{2} \right) = c,$$

for every deg. of inclination of the slope of the embankment from 15 deg. to 40 deg., so that by multiplying $\dfrac{W H^3}{3}$ by this value, double the moment of the pressure of the embankment will be given, and Table No. 5 gives double the moments of different kinds of material accordingly.

In the case of a vertical wall, the moment of its weight $= \dfrac{W H B^2}{2}$, W being the weight of a cubic foot of the wall. Then for equilibrium,

$$\frac{W H B^2}{2} = \frac{c W H^3}{6}, \text{ and B} = .5773\, H \frac{\sqrt{c W}}{\sqrt{W}},$$

and for stability,

$$\frac{W H B^2}{2} = \frac{c W H^3}{3}, \text{ and B} = .81649\, H \frac{\sqrt{c W}}{\sqrt{W}}.$$

Table No. 4 is calculated from the formula $.81649\, H \sqrt{c\,W}$, so that the figures in that Table, divided by the square root of the weight of a cubic foot of the wall, will give the thickness of the wall required for stability.

Table No. 5 gives double the moments of the pressure of the weight of different materials to form a surcharged embankment, with a retaining wall up to 30 ft. in height, and which if made equal to either of the moments of the weight of different forms of retaining walls given afterwards, the dimensions required for that form of wall can be at once found.

The moment of a wall of this section is

$$\frac{W\,H}{2}\left((B+S)^2 - \frac{S^2}{3}\right),$$

where B is the vertical portion of the wall,

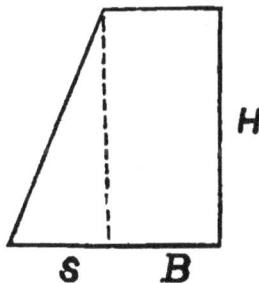

and S is the slope. If $S = \frac{1}{4}$, or 3" to a ft., its moment

$$= \frac{W\,H}{2}\left(\left(B \times \frac{H}{4}\right)^2 - \frac{H^2}{48}\right).$$

The moment of a battering wall of equal thickness

$$= \frac{W\,H\,B}{2}\,(B + S\,H),$$

where B = thickness of wall, and S H = the batter of the slope on the face. If

$$S = \frac{1}{4}, \text{ its moment} = \frac{W H B}{2}\left(\cdot B + \frac{H}{4}\right),$$

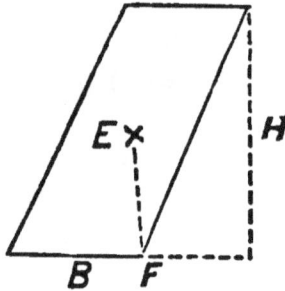

and if E F, the perpendicular from its centre of gravity, falls on its inside corner, its moment = W H B², and the wall then will have the greatest amount of resisting power with security, and also with a minimum amount of material in it. In that case, if M = moment of earth, W = weight of a cubic foot of the wall; for stability,

$$S = \sqrt{\frac{2 M}{W H^3}}, \text{ S H being} = B.$$

To exemplify this, let H = 20 feet, $S = \frac{1}{4}$, W = sand of 120 lbs. to the cubic foot in a surcharged embankment, W = brick of 120 lbs. to the cubic foot in the wall. Then by Table No. 5, the double moment of that kind of sand = 160,000. Then for the first section of wall,

$$\frac{120 \times 20}{2}\left(\left(B+\frac{20}{4}\right)^2 - \frac{20^2}{48}\right) = 160,000, \text{ and}$$

B = 6.9. In this case, weight of wall

$$= 120\left((20 \times 6.9) + \left(\frac{5 \times 20}{2}\right)\right) = 22,560.$$

For second section of wall,

$$\frac{B \times 20 \times 120}{2}\left(B+\frac{20}{4}\right) = 160,000,$$

and B = 9.31, weight of wall = 9.31 $\times$ 20 $\times$ 120 = 22344. For second section of wall, and a perpendicular from its centre of gravity to fall on its inside corner, 120 $\times$ 20 $\times$ B^2 = 160,000, and B = 8.16, weight of wall = 8.16 $\times$ 120 $\times$ 20 = 19593 only, showing a considerable saving of material with this wall.

At the same time, though this wall has the greatest amount of resisting power with the smallest amount of material in it, yet perhaps it may be a question if it would not be advisable to make walls of great height thicker from their base upwards to one-third of their height, which is the centre of pressure.

If we now consider a wall of this form of cross-section, the outside slope of which is

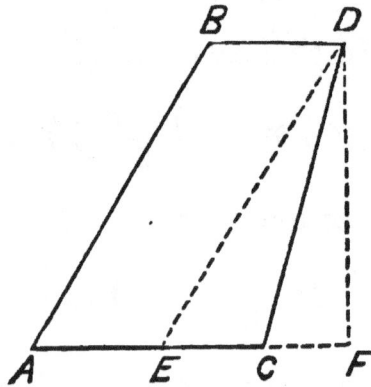

S to 1, and the inside slope next to the embankment S' to 1, we find that its weight is

$$W H B + \frac{W H^2}{2} (S - S'),$$

and the moment of its weight

$$= \frac{W H}{2} \left(H (S - S') \left(\frac{2 S H - S' H}{3} + B \right) + B (S H + B) \right),$$

or if we call it C E and C F, where C E is the difference between the slopes of the front and back of the wall, D E being drawn parallel to the face A B, and C F is the batter of the back of the wall, then its weight is

$$= W H \left(B + \frac{E C}{2} \right),$$

and the moment of its weight

$$= W H \left(B \left(\frac{E F + B}{2} \right) + E C \left(\frac{E C}{3} + \frac{C F}{6} + \frac{B}{2} \right) \right).$$

Then, if the height of the wall be 20 ft., and its weight be 120 lbs. per cubic foot, as before, its outside slope $\frac{1}{4}$ to 1, and its inside slope next to the embankment $\frac{1}{8}$ to 1, then $C F = 2\frac{1}{2}$ ft., $E C = 2\frac{1}{2}$ ft., and its moment

$$= 120 \times 20 \left(B \left(\frac{5 + B}{2} \right) + 2\frac{1}{2} \left(\frac{2\frac{1}{2}}{3} + \frac{2\frac{1}{2}}{6} + \frac{B}{2} \right) \right).$$

$= 160,000$, the double moment of the embankment.

From this equation we find $B = 8.088$, and therefore the weight of that wall

$$= 120 \times 20 \left(8.088 + \frac{2.5}{2} \right) = 22411,$$

and which is, what might have been expected from the form of its cross-section,

being between that of the first form of wall
before mentioned, whose weight was 22560,
and that of the second form, whose weight
was 22344, less than the one and more than
the other.

The form of cross-section of wall, having
its front and back parallel, with the perpendicular from its centre of gravity falling on
its inside corner, having been proved to be
the most economical in material, it may be
asked, why should not this principle be car-
ried further, and walls generally be built
thicker at the top than at the bottom, so as
to have their centre of gravity higher up?
This, by increasing the distance of a perpendicular from it to the outside edge of
the wall at its foot, would much increase its
resisting power to the overturning force of
the bank. It no doubt could be done, and
where the wall is of great thickness it may
be safe to do so, but as there is a fear, how-
ever, of too much reducing the thickness of
the wall at one-third of its height, where is
the centre of pressure, perhaps it may be
advisable to make the form of equal thickness throughout, the limit of our endeavor

to economize material with these forms of wall.

The moment of this form of wall, with its vertical side against the embankment, is

$$\frac{W\,H\,B^2}{3},$$

and if it be required to support water, whose

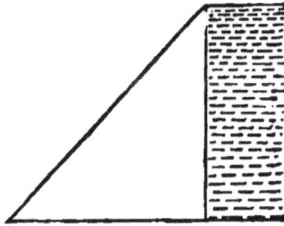

double moment is $20.83\,H^3$, we find from the equation

$$\frac{W\,H\,B^2}{3} = 20.83\,H^3, \quad B = \frac{7.9\,H}{\sqrt{W}},$$

W being the weight of a cubic foot of the wall.

When the sloping side of the wall is next to the water, the pressure of the water on it assists the resisting power of the wall. Its moment is

$$\frac{W\,H\,B^2}{6},$$

and the pressure of the water on the slope

$$S = 62.5\ S \times \frac{H}{2} = 31.25\ S\,H.$$

Thus, when resolved into the horizontal and vertical forces, the former is

$$31.25 \, S \, H \times \sin. \angle a = 31.25 \, S \, H \times \frac{H}{S} = 31.25 \, H^2,$$

and the latter is

$$31.25 \, S \, H \times \cos. a = 31.25 \, S \, H \times \frac{B}{S} = 31.25 \, HB.$$

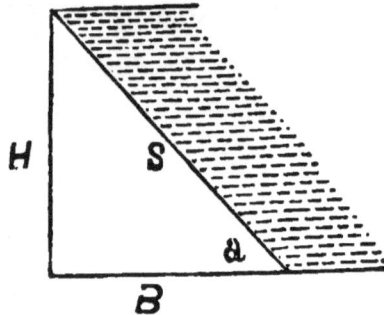

The moment of the former force

$$= 31.25 \, H^2 \times \frac{H}{3} = 10.41\dot{6} \, H^3,$$

and which tends to overturn the wall; and the moment of the latter force

$$= 31.25 \, H \, B \times \frac{2 \, B}{3} = 20.8\dot{3} \, H \, B^2,$$

and which tends to assist the wall. The total moment of the wall for stability must therefore $= 2$ (moment horizontal force — moment vertical force)

$$= 2 \, (10.41\dot{6} \, H^3 - 20.8\dot{3} \, H \, B^2) = $$
$$20.8\dot{3} \, H \, (H^2 - 2 \, B^2).$$

Then

$$\frac{W\,H\,B^2}{6} = 20.8\dot{3}\,H\,(H^2 - 2\,B^2),$$

and

$$B = \frac{11\,18\,H}{\sqrt{W + 250}}.$$

If we take $H = 20$ feet, and $W = 120$ lbs. per cubic foot, then in the first case,

$$B = \frac{7.9 \times 20}{\sqrt{120}} = 14.42,$$

and the weight of the wall

$$= \frac{120 \times 20 \times 14.42}{2} = 17304;$$

and in the second case

$$B = \frac{11.18 \times 20}{\sqrt{120 + 250}}\ 11.62,$$

and the weight of the wall

$$= \frac{120 \times 20 \times 11.62}{2} = 13949.$$

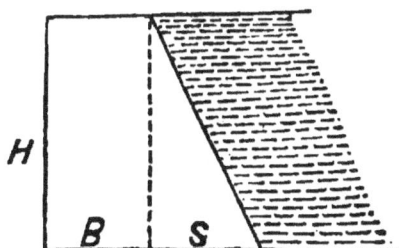

The moment of a wall of this section is

$$\frac{W\,H}{2}\left((B + S)^2 - \frac{S^2}{3}\right),$$

as before mentioned, when the water presses against the vertical side, but if it is on the slope, the moment is

$$\frac{W\,H}{2}\left(B\,(B+S)+\frac{S^2}{3}\right).$$

If we have an embankment of this form of

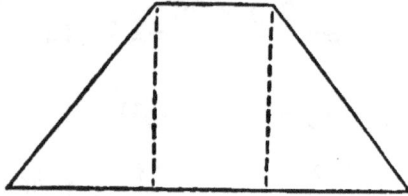

cross-section, where the slopes are the same on both sides, its moment is

$$W\,H\left(B\left(\frac{B+3\,S}{2}\right)+S^2\right).$$

If the steeper slope is on the inside of the embankment, its moment is

$$W\,H\left(B\left(S+\frac{B+S'}{2}\right)+\left(S+\frac{S'}{2}\right)\left(\frac{S+S'}{3}\right)\right).$$

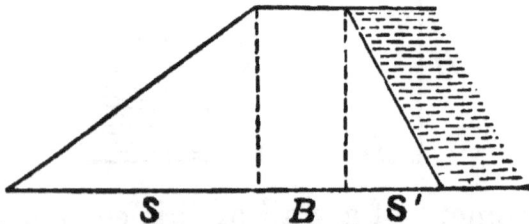

If the steeper slope is on the outside of the embankment, its moment is

$$W H \left(B \left(S' + \frac{B+S}{2} \right) + \left(S' + \frac{S}{2} \right) \left(\frac{S'+S}{3} \right) \right).$$

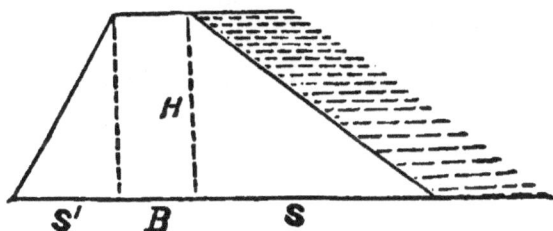

If in these last five equations $W = 120$ lbs. to the cubic foot, $H = 20$ feet, $S = 20$ feet, $S' = 10$ feet, and $B = 10$ feet, then the moment of the first section

$$= \frac{120 \times 20}{2} \left((10 + 20)^2 - \frac{20^2}{3} \right) = 920,000 ;$$

of the second section

$$= \frac{120 \times 20}{2} \left(10 \, (10 + 20) + \frac{20^2}{3} \right) = 520,000 ;$$

of the third section

$$= 120 \times 20 \left(10 \left(\frac{10 + 3 \times 20}{2} \right) + 20^2 \right)$$
$$= 1,800,000 ;$$

of the fourth section

$$= 120 \times 20 \left(10 \left(20 + \frac{10 + 10}{2} \right) + \left(20 + \frac{10}{2} \right) \right.$$
$$\left. \left(\frac{20 + 10}{3} \right) \right) = 1,320,000 ;$$

of the fifth section

$$= 120 \times 20 \left(10 \left(10 + \frac{10 + 20}{2} \right) + \left(10 + \frac{20}{2} \right) \left(\frac{10 + 20}{3} \right) \right) = 1,080,000.$$

In these equations the moments of the walls are to be made equal to twice the difference of the moments of the horizontal and vertical forces of the water, as before, when the sloping side is next to the water. If the wall is to be built with a curved batter instead of a slope, to facilitate the calculation of its moment we may assume the curve to be of a parabolic form, and from which, in the curves generally used for that purpose, it will not sensibly differ. The calculations of the moments of a few forms of wall with curved batter are given, to show how they have been arrived at.

To find the moment of a retaining wall with curved batter generally, let A B E be of the parabolic form, then the area of

$$A\,B\,E = \frac{H}{3} \times B\,E.$$

Now the centre of gravity of A B E will be found sufficiently correct for all practical purposes if it is taken to be in the perpendicular line G F, which will bisect A B E.

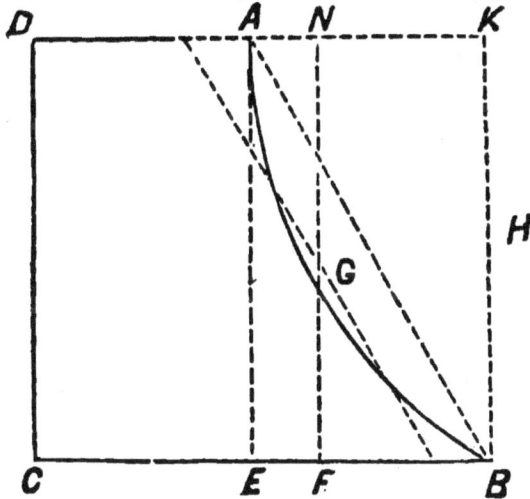

Now

$$A E F G = A E F N - A G N =$$

$$H \times E F - \frac{2}{3} G N \times E F = \frac{A B E}{2} = \frac{H}{6} \times B E,$$

$$A K = B E : B K^2 - H^2 : : A N = E F : G N^2,$$

$$G N = H \sqrt{\frac{E F}{B E}}$$

$$H \times E F - \frac{2}{3} H \sqrt{\frac{E F}{B E}} \times E F = \frac{H}{6} B E,$$

$$\frac{2}{3} \sqrt{\frac{E F}{B E}} = 1 - \frac{B E}{6 E F},$$

$$\frac{E F}{B E} = \frac{9}{4} - \frac{3 B E}{4 E F} + \frac{B E^2}{16 E F^2},$$

$$BE^3 - 12\,BE^2 \times EF + 36\,BE \times EF^2 - 16\,EF^3 = 0,$$
$$BE - 4\,EF)\,(BE^2 - 8\,BE \times EF + 4\,EF^2) = 0,$$
$$BE - 4\,EF = 0, \quad BF = \frac{3\,BE}{4}.$$

Moment of

$$ABE = \frac{H}{3}\,BE \times \frac{3\,BE}{4} = H\,\frac{BE^2}{4};$$

moment of

$$AECD = (H \times CE)\left(BE + \frac{CE}{2}\right);$$

moment of

$$ABCD = H\left(\frac{CE^2}{2} + CE \times BE + \frac{BE^2}{4}\right)$$
$$= \frac{H}{2}\left((CE + BE)^2 - \frac{BE^2}{2}\right).$$

If we take a triangle of equal area with A G B E, and similar to a triangle A B E, we shall find that its base will ·

$$= BE\,\sqrt{\frac{2}{3}} = .8165\,BE,$$

and therefore the distance of a perpendicular from its centre of gravity to

$$E = \frac{.8165\,BE}{3} = .2722\,BE,$$

and therefore, $BE - .2722\,BE = .7278\,BE$ from B, or nearly the same as before. Let $CE = 6$, and other values as before, then

$$120 \frac{20}{2} \left((6 + BE)^2 - \frac{BE^2}{2} \right) =$$

$$160{,}00\dot{0}, \ BE = 6.403,$$

and weight of wall

$$= 120 \left(20 \times 6 + \frac{1}{3} \ 20 \times 6.4 \right) = 19523.$$

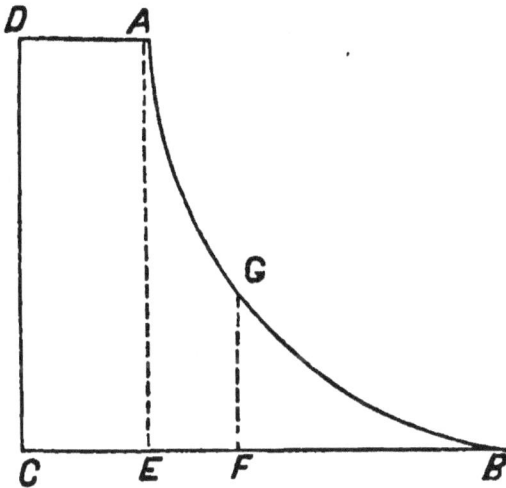

To find the moment of A B C D when A B E = A E C D. Then

$$BE = 3\,CE, \text{ area of } ABE = H \times \frac{BE}{3} = H \times CE.$$

Moment of

$$ABCD = (H \times CE)\left(BE + \frac{CE}{2}\right) +$$

$$\left(H \times \frac{BE}{3}\right)\left(\frac{3\,BE}{4}\right) = (H \times CE)$$

$$\left(3\,C\,E + \frac{C\,E}{2}\right) + (H \times C\,E)\frac{3}{4}\,3\,C\,E - H\frac{7}{2}\,C\,E^2$$

$$+ H\frac{9}{4}\,C\,E^2 = H\frac{23}{4}\,C\,E^2 = H\frac{23}{64}\,B\,C^2.$$

Then

$$120 \times 20\frac{23}{64}\,B\,C^2 = 160,000,$$

$$B\,C = 13.62,\ C\,E = 3.4,\ B\,E = 10.21,$$

and weight of wall,

$$= 120\left(20 \times 3.4 + 20\frac{10.21}{3}\right) = 16344.$$

When both the front and back of the wall are curved and parallel. When E F passes through the centre of gravity, to find E A. Area of

$$E\,A\,B\,F = H \times E\,A + \frac{H}{3}(B\,F - E\,A) =$$

$$\frac{2\,H}{3}\,E\,A + \frac{H}{3}\,B\,F,$$

area of

$$E\,F\,C\,D = H \times C\,F + \frac{2\,H}{3}(E\,D - C\,F)$$

$$= \frac{H}{3}\,C\,F + \frac{2\,H}{3}\,E\,D,$$

then, when E F bisects A B C D,

$$\frac{2\,H}{3}\,E\,A + \frac{H}{3}\,B\,F = \frac{H}{3}\,C\,F + \frac{2\,H}{3}\,E\,D,$$

$$B\,F = C\,F + 2\,(E\,D - E\,A).$$

For stability,

$$\mathrm{B\,F} = \frac{3\,\mathrm{B\,C}}{4} = \frac{3\,\mathrm{A\,D}}{4},$$

$$\mathrm{C\,F} = \frac{\mathrm{B\,C}}{4} = \frac{\mathrm{A\,D}}{4},$$

$$\frac{3\,\mathrm{A\,D}}{4} = \frac{\mathrm{A\,D}}{4} + 2\,(\mathrm{E\,D} - \mathrm{E\,A}),$$

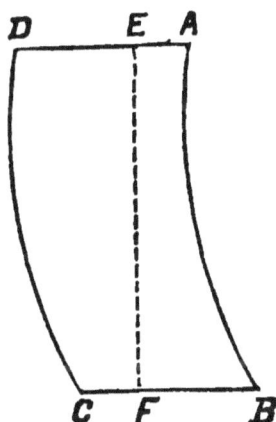

$$\mathrm{A\,D} = 4\,\mathrm{E\,D} - 4\,\mathrm{E\,A},$$

and as

$$4\,\mathrm{A\,D} = 4\,\mathrm{E\,D} + 4\,\mathrm{E\,A},$$

$$3\,\mathrm{A\,D} = 8\,\mathrm{E\,A}, \quad \mathrm{E\,A} = \frac{3}{8}\,\mathrm{A\,D}.$$

To find E A when the perpendicular which bisects A B C D passing through its centre of gravity falls on its inside corner. Area of

$$\mathrm{E\,C\,D} = \frac{2}{3}\,\mathrm{H} \times \mathrm{E\,D} = \frac{2}{3}\,\mathrm{H}\,(\mathrm{A\,D} - \mathrm{E\,A}).$$

Area of

$$\mathbf{A\,B\,C\,E} = \mathbf{H} \times \mathbf{E\,A} + \frac{\mathbf{H}}{3}\,(\mathbf{B\,C} - \mathbf{E\,A}),$$

$$\frac{2\,\mathbf{H}}{3}\,(\mathbf{A\,D} - \mathbf{E\,A}) = \mathbf{H} \times \mathbf{E\,A} + \frac{\mathbf{H}}{3}\,(\mathbf{B\,C} - \mathbf{E\,A}),$$

$$= \frac{2\,\mathbf{H}}{3}\,\mathbf{E\,A} + \frac{\mathbf{H}}{3}\,\mathbf{B\,C},$$

$$\frac{4\,\mathbf{H}}{3}\,\mathbf{E\,A} = \frac{2\,\mathbf{H}}{3}\,\mathbf{A\,D} - \frac{\mathbf{H}}{3}\,\mathbf{B\,C} = \frac{\mathbf{H}}{3}\,\mathbf{B\,C},$$

$$\mathbf{E\,A} = \frac{\mathbf{B\,C}}{4}.$$

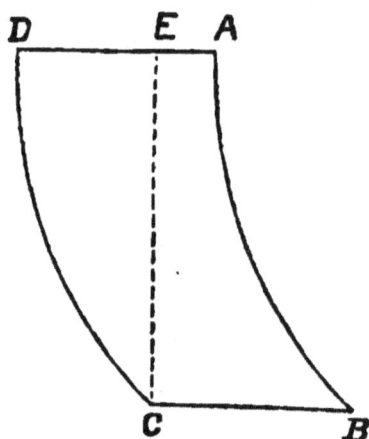

To find the moment of A B C D when the curves of the front and back of the wall are of different radii. Area of

$$\mathbf{E\,F\,C\,D} = \mathbf{H} \times \mathbf{C\,F} + \frac{2\,\mathbf{H}}{3}\,(\mathbf{D\,E} - \mathbf{C\,F})$$

$$= \frac{\mathbf{H}}{3}\,\mathbf{C\,F} + \frac{2\,\mathbf{H}}{3}\,\mathbf{D\,E},$$

area of

$$E\,A\,B\,F = H \times E\,A + \frac{H}{3}\,(B\,F - E\,A)$$

$$= \frac{2\,H}{3}\,E\,A + \frac{H}{3}\,B\,F,$$

area of

$$A\,B\,C\,D = \frac{H}{3}\,C\,F + \frac{2\,H}{3}\,D\,E + \frac{2\,H}{3}\,E\,A +$$

$$\frac{H}{3}\,B\,F = \frac{H}{3}\,(B\,C + 2\,D\,A)\,;$$

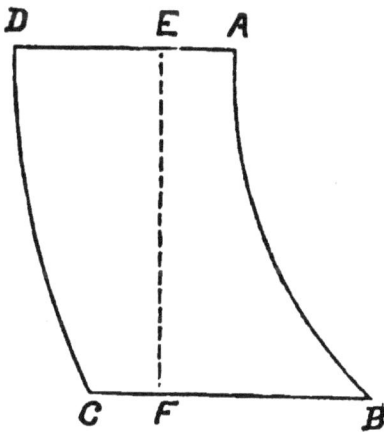

moment of A B C D for stability

$$= \frac{3\,B\,C}{4} \times \frac{H}{3}\,(B\,C + 2\,D\,A)$$

$$= \frac{H \times B\,C}{4}\,(B\,C + 2\,D\,A).$$

As

$$\frac{H}{3}\,C\,F + \frac{2\,H}{3}\,D\,E = \frac{2\,H}{3}\,E\,A + \frac{H}{3}\,B\,F,$$

$$C\,F + 2\,D\,E = 2\,E\,A + B\,F,$$

adding 2 E A to both sides,

$$C F + 2 DA = 4 EA + B F,$$

$$E A = \frac{2 D A + C F - B F}{4}$$

generally, and for stability,

$$E A = \frac{D A}{2} + \frac{\frac{B C}{4}}{4} - \frac{\frac{3 B C}{4}}{4} = \frac{D A}{2} - \frac{B C}{8}.$$

If D A is to be $\frac{3}{4}$ B C, then the moment

$$\frac{H \times B C}{4} (B C + 2 D A)$$

will be

$$\frac{H \times B C}{4} \left(B C + \frac{3}{2} B C \right) = H \frac{5 B C^2}{8}.$$

Then, with values as before,

$$120 \times 20 \frac{5 B C^2}{8} = 160{,}000, \; B C = 10.32,$$

$$D A = \frac{3}{4} \; 10.32 = 7.74,$$

weight of wall

$$= 120 \times \frac{20}{3} (10.32 + 2 \times 7.74) = 20640.$$

If D A is to be $\frac{2}{3}$ B C, then the moment

$$\frac{H \times B C}{4} (B C + 2 D A)$$

will be

$$\frac{H \times BC}{4}\left(BC + \frac{4}{3}BC\right) = H\frac{7\,BC^2}{12}.$$

Then, with values as before,

$$120 \times 20\frac{7\,BC^2}{12} = 160,000,\ BC = 10.69,$$

$$DA = \frac{2}{3}\,10.69 = 7.13,$$

weight of wall

$$= 120 \times \frac{20}{3}(10.69 + 2 \times 7.13) = 19960.$$

If D A is to be $\frac{1}{2}$ B C, then the moment

$$\frac{H \times BC}{4}(BC + 2\,DA)$$

will be

$$\frac{H \times BC}{4}(BC + BC) = H\frac{BC^2}{2}.$$

Then, with values as before,

$$120 \times 20\frac{BC^2}{2} = 160,000,\ BC = 11.54,$$

$$DA = \frac{1}{2}\,11.54 = 5.77,$$

weight of wall

$$= 120 \times \frac{20}{3}(11.54 + 2 \times 5.77) = 18464.$$

If D A is to be $\frac{1}{4}$ B C, then the moment

$$\frac{H \times BC}{4}(BC + 2\,DA)$$

will be

$$\frac{H \times B C}{4}\left(B C + \frac{1}{2} B C\right) = H \frac{3 B C^2}{8}.$$

Then, with values as before,

$$120 \times 20 \frac{3 B C^2}{8} = 160{,}000, \ B C = 13.\dot{3},$$

$$D A = \frac{1}{4} \ 13.\dot{3} = 3.\dot{3},$$

weight of wall

$$= 120 \times \frac{20}{3} (13.\dot{3} + 2 \times 3.\dot{3}) = 16000.$$

If a wall of this section is required, its moment is

$$H \times \frac{B C^2}{4},$$

and if it supports water level with the top,

$$120 \times 20 \frac{B C^2}{4} = 166{,}666, \ B C = 16.\dot{6};$$

and weight of wall

$$= 123 \times 20 \times \frac{16.\dot{6}}{3} = 13333.$$

Now as 17304 was required for the triangular form of wall with the same values, there is shown to be a great saving of material with the form of wall with curved batter.

As the form of wall with a curved batter of the semi-cubical parabolic section, has been proved by several writers to be everywhere of equal strength, the calculations for finding the dimensions of retaining

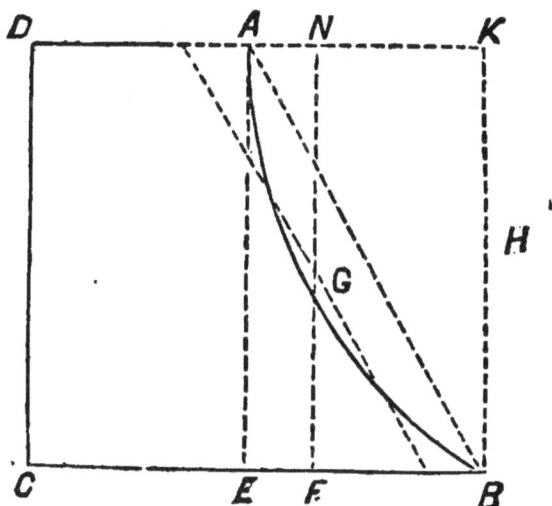

walls with a batter of that curve are also given, as they may be found useful in some cases. Let A G B in this figure be a curve of that form, with G F passing through its centre of gravity. Then

$$AK^2 = BE^2 : BK^3 = H^3 :: AH^2 = EF^2 : GN^3,$$

$$GN = H \sqrt[3]{\frac{EF^2}{BE^2}},$$

area of
$$ABE = \frac{2H}{5} \times BE,$$

area of
$$AGN = \frac{3AN}{5} \times GH,$$

area of
$$AEFG = H \times EF - \frac{3}{5} AN \times GN,$$

$$H \times EF - \frac{3}{5} AN \times NG = \frac{H}{5} \times BE,$$

$$H \times EF - \frac{3}{5} EF \times H \sqrt[3]{\frac{EF^2}{BE^2}} = \frac{H}{5} \times BE,$$

$$EF - \frac{3}{5} EF \sqrt[3]{\frac{EF^2}{BE^2}} = \frac{BE}{5},$$

$$\frac{3}{5} EF \sqrt[3]{\frac{EF^2}{BE^2}} = EF - \frac{BE}{5},$$

$$EF \sqrt[3]{\frac{EF^2}{BE^2}} = \frac{5EF}{3} - \frac{BE}{3},$$

$$\sqrt[3]{\frac{EF^2}{BE^2}} = \frac{5}{3} - \frac{BE}{3EF} = \frac{5 - \frac{BE}{EF}}{3},$$

$$\frac{EF^2}{BE^2} = \frac{\left(5 - \frac{BE}{EF}\right)^3}{27}, \quad \frac{\left(5 - \frac{BE}{EF}\right)^3}{\frac{EF^2}{BE^2}} = 27,$$

$$\frac{B\,E^2}{E\,F^2}\left(5 - \frac{B\,E}{E\,F}\right)^3 = 27, \quad \frac{B\,E}{E\,F} = 3.759,$$

$$B\,E = 3.759\,E\,F = 3.759\,(B\,E - B\,F),$$

$$3.759\,B\,F = 3.759\,B\,E - B\,E,$$

$$B\,F = \frac{2.759\,B\,E}{3.759} = .734\,B\,E.$$

Moment of

$$A\,B\,E = \frac{2\,H}{5}\,B\,E \times .734\,B\,E = .2936\,B\,E^2 \times H;$$

moment of

$$A\,E\,C\,D = (H \times C\,E)\left(B\,E + \frac{C\,E}{2}\right);$$

moment of

$$A\,B\,C\,D = H\left(\frac{C\,E^2}{2} + C\,E \times B\,E + .2936\,B\,E^2\right)$$

$$= \frac{H}{2}\,((C\,E + B\,E)^2 - .4128\,B\,E^2).$$

Let $C\,E = 6$, and other values as before, then

$$120\,\frac{20}{2}\,((6 + B\,E)^2 - .4128\,B\,E^2) = 160,000,$$

$B\,E = 6.22$, and weight of wall

$$= 120\left(20 \times 6 + \frac{2}{5}\,20 \times 6.22\right) = 20369.$$

To find the moment of A B C D when

$A B E = A E C D.$ Then $B E = \frac{5}{2} C E,$
area of $A B E = H \times \frac{2}{5} B E = H \times C E.$

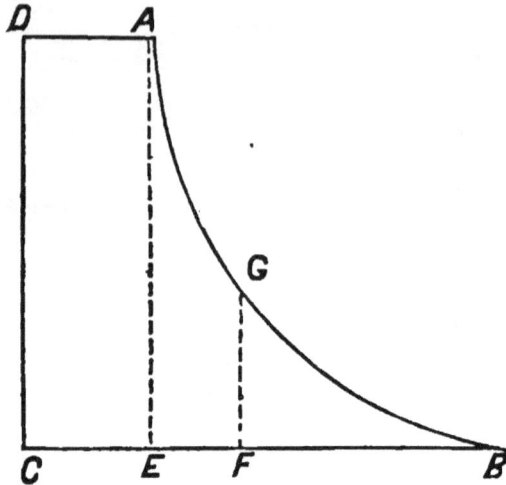

Moment of

$$A B C D = (H \times C E)\left(B E + \frac{C E}{2}\right) +$$

$$\left(H \times \frac{2}{5} B E\right)(.734 \, B E) = (H \times C E)$$

$$\left(\frac{5}{2} C E + \frac{C E}{2}\right) + (H \times C E)\left(.734 \times \frac{5}{2} C E\right)$$

$$= (H \times C E) \, 3 \, C E + (H \times C E) \, 1.835 \, C E$$

$$= 4.835 \, H \times C E^2. \text{ Then } 120 \times 20 \times 4.835 \, C E^2$$

$$= 160,000, \; C E = 3.71, \; B E = \frac{5}{2} \, 3.71 = 9.28,$$

and weight of wall

$$= 120 \times 20 \left(3.71 + \frac{2}{5} \, 9.28\right) = 17808.$$

When both the front and back of the
wall are curved and parallel. When E F
passes through the centre of gravity, to find
E A. Area of

$$E A B F = H \times E A + \frac{2H}{5}(B F - E A)$$

$$= \frac{3H}{5} E A + \frac{2H}{5} B F,$$

area of

$$E F C D = H \times C F + \frac{3H}{5}(E D - C F)$$

$$= \frac{2H}{5} C F + \frac{3H}{5} E D,$$

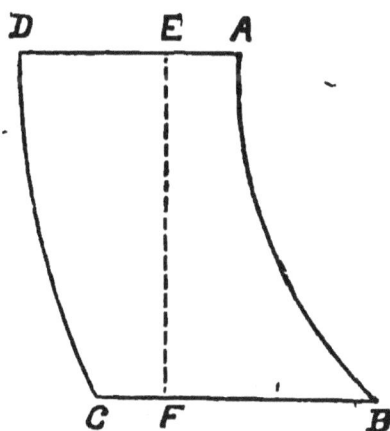

then when E F bisects A B C D,

$$\frac{3H}{5} E A + \frac{2H}{5} B F = \frac{2H}{5} C F + \frac{3H}{5} E D,$$

$$2 B F = 2 C F + 3 E D - 3 E A;$$

for stability,

$$BF = \frac{3\,BC}{4} = \frac{3\,AD}{4}, \; CF = \frac{AD}{4},$$

so

$$\frac{3\,AD}{2} = \frac{AD}{2} + 3\,ED - 3\,EA,$$

$AD = 3\,ED - 3\,EA$, and as $3\,AD = 3\,ED + EA$,

subtracting,

$$2\,AD = 6\,EA, \; EA = \frac{AD}{3}.$$

To find E A when the perpendicular which bisects A B C D passing through its centre of gravity, falls on its inside corner. Area of

$$ECD = \frac{3\,H}{5}\,ED = \frac{3\,H}{5}\,(AD - EA),$$

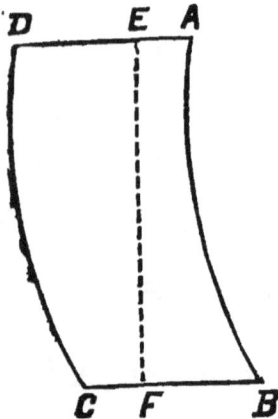

area of

$$ABCE = H \times EA + \frac{2\,H}{5}\,(BC - EA),$$

$$\frac{3\,H}{5}\,(A\,D - E\,A) = H \times E\,A + \frac{2\,H}{5}\,(B\,C - E\,A)$$

$$= \frac{3\,H}{5}\,E\,A + \frac{2\,H}{5}\,B\,C,$$

$$\frac{6\,H}{5}\,E\,A = \frac{3\,H}{5}\,A\,D - \frac{2\,H}{5}\,B\,C = \frac{H}{5}\,B\,C,$$

$$E\,A = \frac{B\,C}{6}.$$

To find the moment of A B C D when the curves of the front and back of the wall are of different radii. Area of

$$E\,F\,C\,D = H \times C\,F + \frac{3\,H}{5}\,(D\,E - C\,F)$$

$$= \frac{2\,H}{5}\,C\,F + \frac{3\,H}{5}\,D\,E,$$

area of

$$E\,A\,B\,F = H \times E\,A + \frac{2\,H}{5}\,(B\,F - E\,A)$$

$$= \frac{3\,H}{5}\,E\,A + \frac{2\,H}{5}\,B\,F,$$

area of

$$A\,B\,C\,D = \frac{2\,H}{6}\,C\,F + \frac{3\,H}{5}\,D\,E +$$

$$\frac{3\,H}{5}\,E\,A + \frac{2\,H}{5}\,B\,F = \frac{H}{5}\,(2\,B\,C + 3\,A\,D);$$

moment of A B C D for stability

$$= \frac{3\cdot B\,C}{4} \times \frac{H}{5}\,(2\,B\,C + 3\,A\,D)$$

$$= \frac{3\,H \times B\,C}{20}\,(2\,B\,C + 3\,A\,D).$$

As $\dfrac{2\,H}{5}\,C\,F + \dfrac{3\,H}{5}\,D\,E = \dfrac{3\,H}{5}\,.\,E\,A + \dfrac{2\,H}{5}\,B\,E,$

$$2\,C\,F + 3\,D\,E = 3\,E\,A + 2\,B\,F,$$

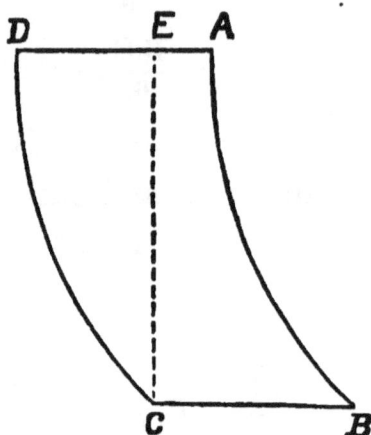

adding $3\,E\,A$ to both sides,

$$2\,C\,F + 3\,D\,A = 6\,E\,A + 2\,B\,F,$$

$$E\,A = \frac{3\,D\,A + 2\,C\,F - 2\,B\,F}{6}$$

generally, and for stability,

$$E\,A = \frac{3\,D\,A + \dfrac{B\,C}{2} - \dfrac{3\,B\,C}{2}}{6} = \frac{D\,A}{2} - \frac{B\,C}{6}.$$

If $D\,A$ is to be $\dfrac{3}{4}\,B\,C$, then the moment

$$\frac{3\,H \times B\,C}{20}\,(2\,B\,C + 3\,A\,D)$$

will be

$$\frac{3\,H \times B\,C}{20}\left(2\,B\,C + \frac{9}{4}\,B\,C\right) = \frac{H\,51\,B\,C^2}{80}.$$

Then, with values as before,

$$120 \times 20\,\frac{51\,B\,C^2}{80} = 160{,}000,\ B\,C = 10.23,$$

$$D\,A = \frac{3}{4}\,10.23 = 7.67,$$

weight of wall

$$= 120 \times \frac{20}{5}\,(2 \times 10.23 + 3 \times 7.67) = 20860.$$

If D A is to be $\frac{2}{3}$ B C, then the moment

$$\frac{3\,H \times B\,C}{20}\,(2\,B\,C + 3\,A\,D)$$

will be

$$\frac{3\,H \times B\,C}{20}\,(2\,B\,C + 2\,B\,C) = H\,\frac{3\,B\,C^2}{5}.$$

Then, with values as before,

$$120 \times 20\,\frac{3\,B\,C^2}{5} = 160{,}000,$$

$$B\,C = 10.54,\ D\,A = \frac{2}{3}\,10.54 = 7.03,$$

weight of wall

$$= 120 \times \frac{20}{5}\,(2 \times 10.54 + 3 \times 7.03) = 20238.$$

If D A is to be $\frac{1}{2}$ B C, then the moment

$$\frac{3\,H \times B\,C}{20}\,(2\,B\,C + 3\,A\,D)$$

will be

$$\frac{3\,H \times B\,C}{20}\left(2\,B\,C + \frac{3}{2}\,B\,C\right) = H\,\frac{21\,B\,C^2}{40}.$$

Then, with values as before,

$$120 \times 20\,\frac{21\,B\,C^2}{40} = 160,000,$$

$$B\,C = 11.27,\ D\,A = \frac{1}{2}\,11.27 = 5.63,$$

weight of wall

$$= 120 \times \frac{20}{5}\,(2 \times 11.27 + 3 \times 5.63) = 18930.$$

If $A\,D$ is to be $\frac{1}{3}\,B\,C$, then the moment

$$\frac{3\,H \times B\,C}{20}\,(2\,B\,C + 3\,A\,D)$$

will be

$$\frac{3\,H \times B\,C}{20}\,(2\,B\,C + B\,C) = H\,\frac{9\,B\,C^2}{20}.$$

Then, with values as before,

$$120 \times 20\,\frac{9\,B\,C^2}{20} = 160,000,$$

$$B\,C = 12.17,\ D\,A = \frac{1}{3}\,12.17 = 4.06,$$

weight of wall

$$120 \times \frac{20}{5}\,(2 \times 12.17 + 3 \times 4.06) = 17526.$$

If D A is to be $\frac{1}{4}$ B C, then the moment

$$\frac{3\,H \times B\,C}{20}\,(2\,B\,C + 3\,A\,D)$$

will be

$$\frac{3\,H \times B\,C}{20}\left(2\,B\,C + \frac{3}{4}\,B\,C\right) = H\,\frac{33\,B\,C^2}{80}.$$

Then, with values as before,

$$120 \times 20\,\frac{33\,B\,C^2}{80} = 160{,}000,$$

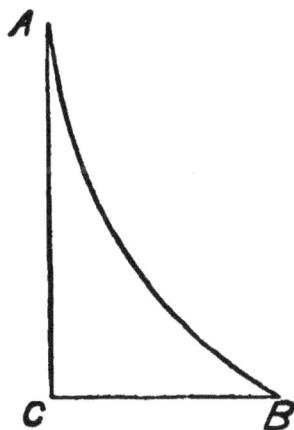

$$B\,C = 12.71,\ D\,A = \frac{1}{4}\,12.71 = 3.18,$$

weight of wall

$$= 120 \times \frac{20}{5}\,(2 \times 12.71 + 3 \times 3.18) = 16780.$$

If a wall of this section is required, its

moment is $.2936$ B $C^2 \times$ H, and if it supports water level with the top,

$$120 \times 20 \times .2936 \text{ B } C^2 = 166,666, \text{ B C} = 15.38,$$

and weight of wall

$$= 120 + 20 \times \frac{2 \times 15.38}{5} = 14764.$$

Having now given methods for finding the correct dimensions of the different forms of wall that are generally used in practice, the author does not wish to express any opinion on the merits of any particular form of wall, leaving it to the superior judgment of more experienced engineers to determine the section of wall they may consider most suitable in each case.

TABLE 1.—*Thickness of Vertical Retaining Walls, to sustain the Pressure of Earth, Sand, etc., level with its top. The Moment of the Wall is equal to twice that of the Earth, etc., to insure permanent stability.*

Height of wall	Sand. $\angle = 30°$.		Shingle. $\angle = 40°$.		Dry earth. $\angle = 43°$.
	94 lbs.	120 lbs.	119 lbs.	106 lbs.	94 lbs.
6	27.42	30.98	24.92	23.52	20.65
7	31.99	36.15	29.07	27.44	24.09
8	36.56	41.31	33.23	31.36	27.53
9	41.13	46.47	37.38	35.28	30.98
10	45.70	51.64	41.53	39.20	34.42
11	50.27	56.80	45.69	43.12	37.86
12	54.84	61.97	49.84	47.04	41.30
13	59.42	67.13	53.99	50.96	44.74
14	63.99	72.29	58.15	54.88	48.19
15	68.56	77.46	62.30	58.80	51.63
16	73.13	82.62	66.45	62.72	55.07
17	77.70	87.79	70.61	66.64	58.51
18	82.27	92.95	74.76	70.56	61.95
19	86.84	98.11	78.91	74.48	65.40
20	91.41	103.28	83.07	78.40	68.84
21	95.98	108.44	87.22	82.32	72.28
22	100.55	113.61	91.38	86.24	75.72
23	105.12	118.77	95.53	90.16	79.17
24	109.69	123.94	99.68	94.08	82.61
25	114.26	129.10	103.84	98.00	86.05
26	118.83	134.26	107.99	101.92	89.49
27	123.40	139.43	112.14	105.84	92.93
28	127.97	144.59	116.30	109.76	96.38
29	132.54	149.76	120.45	113.68	99.82
30	137.11	154.92	124.60	117.60	103.26

TABLE 1.—*Continued.*

Do., moist or natural. $\angle = 54°$.	Do., dense and compact. $\angle = 55°$.	Clay. $\angle = 16°$.	Clay. $\angle = 45°$.
106 lbs.	125 lbs.	125 lbs.	125 lbs.
16.39	17.27	41.27	22.69
19.12	20.15	48.15	26.47
21.85	23.03	55.03	30.25
24.58	25.90	61.91	34.03
27.31	28.78	68.79	37.81
30.04	31.66	75.67	41.59
32.78	34.54	82.55	45.38
35.51	37.42	89.43	49.16
38.24	40.29	96 31	52.94
40.97	43.17	103.18	56.72
43.70	46.05	110.06	60.50
46.43	48.93	116.94	64.28
49.16	51.81	123.82	68.06
51.89	54.69	130.70	71.84
54.63	57.56	137.58	75.62
57.36	60.44	144.46	79.40
60.09	63.32	151.34	83.18
62.82	66.19	158.22	86.96
65.56	69.07	165.10	90.74
68.29	71.95	171.97	94.52
71.02	74.83	178.85	98.31
73.75	77.71	185.73	102.09
76.48	80.58	192.61	105.87
79.21	83.46	199.49	109.65
81.94	86.34	206.37	113.43

TABLE 2.—*Double Moments of the Pressure of the Weight of Embankments of Earth, Sand, etc., level with the top of Wall.*

	Sand.		Shingle.		Dry earth.
$c\dfrac{WH^3}{3}$	$=10.4\,H^3$	$13.3\,H^3$	$8.62522\,H^3$	$7.6829\,H^3$	$5.92394\,H^3$
6	2256	2880	1863	1659	1280
7	3582	4573	2958	2635	2032
8	5347	6827	4416	3934	3033
9	7614	9720	6287	5601	4318
10	10444	13333	8625	7683	5924
11	13901	17747	11480	10226	7885
12	18048	23040	14904	13276	10236
13	22946	29293	18949	16879	13015
14	28659	36587	23667	21081	16255
15	35250	45000	29110	25929	19993
16	42780	54613	35329	31468	24264
17	51313	65507	42376	37745	29104
18	60912	77766	50302	44805	34548
19	71638	91453	59160	52696	40632
20	83555	106666	69002	61461	47391
21	96726	123480	79878	71149	54862
22	111212	141973	91841	81805	63078
23	127077	162227	104943	93475	72077
24	144384	184320	119235	106206	81892
25	163194	208333	134769	121042	92561
26	183571	234346	151597	135035	104119
27	205578	262440	169770	151222	116601
28	229276	292693	189341	168655	130042
29	254729	325186	210360	187378	144479
30	282000	360000	232881	207438	159946

TABLE 2.—*Continued.*

Do., moist or natural.	Do., dense and compact.	Clay.		Water.
$3.73024\ H^3$	$4.14222\ H^3$	$23.66012H^3$	$7.14887\ H^3$	$20.83\ H^3$
806	895	5110	1544	4500
1279	1421	8115	2452	7146
1910	2121	12114	3660	10666
2719	3020	17248	5211	15187
3730	4142	23660	7149	20833
4965	5513	31492	9515	27729
6446	7158	40885	12353	36000
8195	9100	51981	15706	45771
10236	11366	64923	19616	57166
12590	13980	79853	24127	70312
15279	16966	96912	29282	85333
18327	20251	116242	35122	102354
21755	24157	137986	41692	121500
25586	28411	162285	49034	142896
29842	33138	189281	57191	166666
34546	38361	219116	66206	192937
39720	44106	251933	76121	221833
45386	50398	287873	86980	253479
51567	57262	327077	98826	288000
58285	64722	369689	111701	325521
65563	72804	415850	125650	366166
73422	81531	465702	140711	410062
81886	90930	519387	156932	457333
90977	101025	577046	174354	508104
100716	111840	638823	193019	562499

TABLE 3.—*For Surcharged Embankments.*

∠ of slope = ϕ.	$\angle\,\mathrm{A\,B\,C} = \dfrac{90^\circ - \phi}{2}$	Tang. of $\dfrac{90^\circ - \phi}{2}$
4 to 1 = 14° 12′	37° 54′	.77847
15 0	37 30	.76732
16 0	37 0	.75355
17 0	36 30	.73996
18 0	36 0	.72654
3 to 1 = 18 25	35 47½	.72100
19 0	35 30	.71329
20 0	35 0	.70020
21 0	34 30	.68728
22 0	34 0	.67450
23 0	33 30	.66188
24 0	33 0	.64940
25 0	32 30	.63707
26 0	32 0	.62486
2 to 1 = 26 35	31 42½	.61781
27 0	31 30	.61280
28 0	31 0	.60086
29 0	30 30	.58904
1¾ to 1 = 29 44	30 8	.58045
30 0	30 0	.57735
31 0	29 30	.56577
32 0	29 0	.55430
33 0	28 30	.54295
1½ to 1 = 33 42	28 9	.53507
34 0	28 0	.53170
35 0	27 30	.52056
36 0	27 0	.50952
37 0	26 30	.49858
38 0	26 0	.48773
1¼ to 1 = 38 40	25 40	.48055
39 0	25 30	.47697
40 0	25 0	.46630

TABLE 3.—*Continued.*

$\theta = \dfrac{\sin.\ (90° + \phi)}{\sin.\ \left(\dfrac{90° - \phi}{2}\right)}$	$\sqrt{1 - \dfrac{\theta^2}{4}}$	Tang. of $\dfrac{90° - \phi}{2}$ $\left(\theta\ \sqrt{1 - \dfrac{\theta^2}{4}}\right)$
		c
1.5782	0.96944	0.75469
1.5867	0.96592	0.74118
1.5972	0.96126	0.72436
1.6077	0.95630	0.70762
1.6180	0.95105	0.69098
1.6223	0.94878	0.68407
1.6282	0.94551	0.67443
1.6383	0.93969	0.65798
1.6483	0.93358	0.64163
1.6581	0.92718	0.62539
1.6678	0.92050	0.60926
1.6773	0.91354	0.59326
1.6868	0.90630	0.57738
1.6961	0.89879	0.56162
1.7014	0.89428	0.55250
1.7053	0.89100	0.54601
1.7143	0.88294	0.53052
1.7232	0.87461	0.51519
1.7297	0.86834	0.50403
1.7320	0.86602	0.50000
1.7407	0.85716	0.48496
1.7492	0.84804	0.47008
1.7576	0.83867	0.45536
1.7634	0.83195	0.44515
1.7659	0.82903	0.44080
1.7740	0.81915	0.42642
1.7820	0.80901	0.41221
1.7899	0.79863	0.39818
1.7976	0.78801	0.38433
1.8026	0.78079	0.37521
1.80517	0.77714	0.37067
1.81261	0.76604	0.35721

TABLE 4.—*Thickness of Vertical Retaining Walls to sustain the Pressure of a Surcharged Embankment of Earth, Sand, etc. The moment of the Wall is equal to twice that of the Earth, etc., to insure permanent stability.*

Height of wall.	Sand. $\angle = 30°$. $c = 0.500$		Shingle. $\angle = 40°$. $c = 0.35721$		Dry earth. $\angle = 43°$. $c = .318001$
	94 lbs.	120 lbs.	119 lbs.	106 lbs.	94 lbs.
6	33.58	37.94	31.94	30.14	26.78
7	39.18	44.27	37.26	35.17	31.24
8	44 78	50.59	42.58	40.19	35.71
9	50.37	56.92	47.91	45 21	40.17
10	55.97	63.24	53.23	50 24	44.64
11	61 57	69.57	58.55	55.26	49.10
12	67.17	75.89	63.88	60.29	53.56
13	72.76	82.21	69.20	65.31	58.03
14	78.36	88.54	74.52	70.33	62.49
15	83.96	94.86	79.85	75.36	66.96
16	89.56	101.19	85.17	80.38	71.42
17	95.16	107.51	90.49	85.41	75.89
18	100.75	113.84	95.82	90.43	80 35
19	106.35	120.16	101.14	95.46	84.81
20	111.95	126.49	106.46	100.48	89.28
21	117.55	132.81	111.79	105 50	93.74
22	123.14	139.14	117.11	110.53	98.21
23	128.74	145.46	122.43	115.55	102.67
24	134 34	151.78	127.76	120.58	107.13
25	139.94	158.11	133.08	125.60	111.60
26	145.54	164.44	138.41	130.63	116.07
27	151.13	170.76	143.73	135.65	120.53
28	156.73	177.09	149.05	140.68	124.99
29	162.33	183.41	154 38	145.70	129.46
30	167.93	189.74	159.70	150.73	133.92

TABLE 4.—*Continued.*

Do., moist or natural. $\angle = 54°$. $c = .190983$	Do., dense and compact. $\angle = 55°$: $c = .1808438$	Clay. $\angle = 16°$.	Clay. $\angle = 45°$.
106 lbs.	125 lbs.	125 lbs.	125 lbs.
22.04	23.29	46.61	29.64
25.71	27.17	54.38	34.58
29.39	31.05	62.15	39.52
33.06	34.93	69.92	44.46
36.73	38.82	77.69	49.40
40.41	42.70	85.46	54.34
44.08	46.58	93.23	59.28
47.75	50.46	101.00	64.22
51.43	54.34	108.77	69.16
55.10	58.23	116.54	74.10
58.78	62.11	124.31	79.04
62.45	65.99	132.08	83.98
66.12	69.87	139.85	88.92
69.80	73.75	147.62	93.86
73 47	77.64	155.39	98.80
77.14	81.52	163.16	103.74
80.82	85.40	170.93	108.68
84.49	89.28	178.70	113.62
88.16	93.16	186.47	118.56
91.84	97.05	194.24	123.50
95.52	100.93	202.01	128.45
99.19	104.81	209.78	133.39
102 86	108.70	217.55	138.33
106.54	112.58	225.32	143.27
110.21	116.46	233.10	148.21

TABLE 5.—*Double Moments of the Pressure of the Weight of Surcharged Embankments of Earth, Sand, etc.*

$\dfrac{c\,W\,H^3}{3}$	Sand.		Shingle.	
	$=15.6\,H^3$	$20\,H^3$	$14.16945\,H^3$	$12.62153\,H^3$
6	3384	4320	3061	2726
7	5373	6860	4860	4329
8	8021	10240	7255	6462
9	11421	14580	10330	9201
10	15666	20000	14169	12621
11	20852	26620	18860	16799
12	27072	34560	24485	21810
13	34419	43940	31130	27729
14	42989	54880	38881	34633
15	52875	67500	47822	42598
16	64170	81920	58038	51698
17	76970	98260	69614	62010
18	91368	116640	82636	73609
19	107457	137180	97188	86571
20	125333	160000	113355	100972
21	145089	185220	131223	116888
22	166818	212960	150876	134394
23	190616	243340	172400	153566
24	216576	276480	195878	174480
25	244791	312500	221397	197211
26	275357	351520	249042	221836
27	308367	393660	278897	248429
28	343914	439040	311048	277068
29	382094	487780	345579	307826
30	423000	540000	382575	340781

Dry earth.	Do., moist or natural.	Do., dense and compact.	Clay.	
9.96405 H³	6.748066 H³	7 53526 H³	30.183 H³	12.20375 H³
2152	1458	1627	6520	2636
3418	2315	2585	10353	4186
5102	3455	3858	15454	6248
7264	4919	5493	22004	8896
9964	6748	7535	30183	12204
13262	8982	10029	40174	16243
17218	11661	13021	52157	21088
21891	14825	16555	66313	26812
27341	18517	20677	82823	33487
33628	22775	25432	101869	41188
40813	27640	30864	123631	49986
48953	33153	37021	148291	59957
58110	39355	43946	176029	71172
68343	46285	51684	207027	83712
79712	53985	60282	241466	97630
92277	62494	69784	279528	113019
106097	72453	80235	321392	129945
121232	82104	91681	367241	148483
137743	93285	104167	417254	168704
155688	105438	117738	471615	190684
175128	118604	132439	530502	214493
196122	132822	148316	594098	240206
218731	148133	165414	662584	267897
243013	164578	183777	736141	297637
269029	182198	203452	814949	329501

VALUABLE

SCIENTIFIC BOOKS,

PUBLISHED BY

D. VAN NOSTRAND,

23 MURRAY STREET AND 27 WARREN STREET,

NEW YORK.

FRANCIS. Lowell Hydraulic Experiments, being a selection from Experiments on Hydraulic Motors, on the Flow of Water over Weirs, in Open Canals of Uniform Rectangular Section, and through submerged Orifices and diverging Tubes. Made at Lowell, Massachusetts. By James B. Francis, C. E. 2d edition, revised and enlarged, with many new experiments, and illustrated with twenty-three copperplate engravings. 1 vol. 4to, cloth....................$15 00

ROEBLING (J. A.) Long and Short Span Railway Bridges. By John A. Roebling, C. E. Illustrated with large copperplate engravings of plans and views. Imperial folio, cloth............................. 25 00

CLARKE (T. C.) Description of the Iron Railway Bridge over the Mississippi River, at Quincy, Illinois. Thomas Curtis Clarke, Chief Engineer. Illustrated with 21 lithographed plans. 1 vol. 4to, cloth 7 50

TUNNER (P.) A Treatise on Roll-Turning for the Manufacture of Iron. By Peter Tunner. Translated and adapted by John B. Pearse, of the Pennsylvania Steel Works, with numerous engravings wood cuts and folio atlas of plates................$10 00

ISHERWOOD (B. F.) Engineering Precedents for Steam Machinery. Arranged in the most practical and useful manner for Engineers. By B. F. Isherwood, Civil Engineer, U. S. Navy. With Illustrations. Two volumes in one. 8vo, cloth.......... $2 50

BAUERMAN. Treatise on the Metallurgy of Iron, containing outlines of the History of Iron Manufacture, methods of Assay, and analysis of Iron Ores, processes of manufacture of Iron and Steel, etc , etc. By H. Bauerman. First American edition. Revised and enlarged, with an Appendix on the Martin Process for making Steel, from the report of Abram S. Hewitt. Illustrated with numerous wood engravings. 12mo, cloth........................... .... 2 00

CAMPIN on the Construction of Iron Roofs. By Francis Campin. 8vo, with plates, cloth........... 3 00

COLLINS. The Private Book of Useful Alloys and Memoranda for Goldsmiths, Jewellers, &c. By James E. Collins. 18mo, cloth.................. 75

CIPHER AND SECRET LETTER AND TELEGRAPHIC CODE, with Hogg's Improvements. The most perfect secret code ever invented or discovered. Impossible to read without the key. By C. S. Larrabee. 18mo, cloth. 1 00

COLBURN. The Gas Works of London. By Zerah Colburn, C. E. 1 vol 12mo, boards.............. 60

CRAIG (B. F.) Weights and Measures. An account of the Decimal System, with Tables of Conversion for Commercial and Scientific Uses. By B. F. Craig, M.D. 1 vol. square 32mo, limp cloth.... 50

NUGENT. Treatise on Optics; or, Light and Sight, theoretically and practically treated; with the application to Fine Art and Industrial Pursuits. By E. Nugent. With one hundred and three illustrations. 12mo, cloth...................................... 2 00

GLYNN (J.) Treatise on the Power of Water, as applied to drive Flour Mills, and to give motion to Turbines and other Hydrostatic Engines. By Joseph

Glynn. Third edition, revised and enlarged, with numerous illustrations. 12mo, cloth. $1 00

HUMBER. A Handy Book for the Calculation of Strains in Girders and similar Structures, and their Strength, consisting of Formulæ and corresponding Diagrams, with numerous details for practical application. By William Humber. 12mo, fully illustrated, cloth. 2 50

GRUNER. The Manufacture of Steel. By M. L. Gruner. Translated from the French, by Lenox Smith, with an appendix on the Bessamer process in the United States, by the translator. Illustrated by Lithographed drawings and wood cuts. 8vo, cloth. . 3 50

AUCHINCLOSS. Link and Valve Motions Simplified. Illustrated with 37 wood-cuts, and 21 lithographic plates, together with a Travel Scale, and numerous useful Tables. By W. S. Auchincloss. 8vo, cloth. . 3 00

VAN BUREN. Investigations of Formulas, for the strength of the Iron parts of Steam Machinery. By J. D. Van Buren, Jr., C. E. Illustrated, 8vo, cloth. 2 00

JOYNSON. Designing and Construction of Machine Gearing. Illustrated, 8vo, cloth. 2 00

GILLMORE. Coignet Beton and other Artificial Stone. By Q. A. Gillmore, Major U S. Corps Engineers. 9 plates, views, &c. 8vo, cloth. 2 50

SAELTZER. Treatise on Acoustics in connection with Ventilation. By Alexander Saeltzer, Architect. 12mo, cloth. 2 00

THE EARTH'S CRUST. A handy Outline of Geology. By David Page. Illustrated, 18mo, cloth. 75

DICTIONARY of Manufactures, Mining, Machinery, and the Industrial Arts. By George Dodd. 12mo, cloth. 2 00

FRANCIS. On the Strength of Cast-Iron Pillars, with Tables for the use of Engineers, Architects, and Builders. By James B. Francis, Civil Engineer. 1 vol. 8vo, cloth, ... 2 00

3

GILLMORE (Gen. Q. A.) Treatise on Limes, Hydraulic Cements, and Mortars. Papers on Practical Engineering, U. S. Engineer Department, No. 9, containing Reports of numerous Experiments conducted in New York City, during the years 1858 to 1861, inclusive. By Q. A. Gillmore, Bvt. Maj -Gen., U. S. A., Major, Corps of Engineers. With numerous illustrations. 1 vol, 8vo, cloth............... $4 00

HARRISON. The Mechanic's Tool Book, with Practical Rules and Suggestions for Use of Machinists, Iron Workers, and others. By W. B. Harrison, associate editor of the "American Artisan." Illustrated with 44 engravings. 12mo, cloth............ 1 50

HENRICI (Olaus). Skeleton Structures, especially in their application to the Building of Steel and Iron Bridges. By Olaus Henrici. With folding plates and diagrams. 1 vol. 8vo, cloth.................. 3 00

HEWSON (Wm.) Principles and Practice of Embank ing Lands from River Floods, as applied to the Levees of the Mississippi. By William Hewson, Civil Engineer. 1 vol. 8vo, cloth..................... 2 00

HOLLEY (A. L.) Railway Practice. American and European Railway Practice, in the economical Generation of Steam, including the Materials and Construction of Coal-burning Boilers, Combustion, the Variable Blast, Vaporization, Circulation, Superheating, Supplying and Heating Feed-water, etc., and the Adaptation of Wood and Coke-burning Engines to Coal-burning; and in Permanent Way, including Road-bed, Sleepers, Rails, Joint-fastenings, Street Railways, etc., etc. By Alexander L. Holley, B. P. With 77 lithographed plates. 1 vol. folio, cloth.... 12 00

KING (W. H.) Lessons and Practical Notes on Steam, the Steam Engine, Propellers, etc., etc., for Young Marine Engineers, Students, and others. By the late W. H. King, U. S. Navy. Revised by Chief Engineer J. W. King, U. S. Navy. Twelfth edition, enlarged. 8vo, cloth. 2 00

MINIFIE (Wm.) Mechanical Drawing. A Text-Book of Geometrical Drawing for the use of Mechanics

4

an⸱ School⸱, in which the Definitions and Rules of Geometry are familiarly explained; the Practical Problems are arranged, from the most simple to the more complex, and in their description technicalities are avoided as much as possible. With illustrations for Drawing Plans, Sections, and Elevations of Railways and Machinery; an Introduction to Isometrical Drawing, and an Essay on Linear Perspective and Shadows. Illustrated with over 200 diagrams engraved on steel. By Wm. Minifie, Architect. Seventh edition. With an Appendix on the Theory and Application of Colors. 1 vol. 8vo, cloth........... $4 00

"It is the best work on Drawing that we have ever seen, and is especially a text-book of Geometrical Drawing for the use of Mechanics and Schools. No young Mechanic, such as a Machinist, Engineer, Cabinet-maker, Millwright, or Carpenter, should be without it."—*Scientific American.*

———— Geometrical Drawing. Abridged from the octavo edition, for the use of Schools. Illustrated with 48 steel plates. Fifth edition. 1 vol. 12mo, cloth.... 2 00

STILLMAN (Paul.) Steam Engine Indicator, and the Improved Manometer Steam and Vacuum Gauges—their Utility and Application. By Paul Stillman. New edition. 1 vol. 12mo, flexible cloth........... 1 00

SWEET (S. H.) Special Report on Coal; showing its Distribution, Classification, and cost delivered over different routes to various points in the State of New York, and the principal cities on the Atlantic Coast. By S. H. Sweet. With maps, 1 vol. 8vo, cloth..... 3 00

WALKER (W. H.) Screw Propulsion. Notes on Screw Propulsion: its Rise and History. By Capt. W. H. Walker, U. S. Navy. 1 vol. 8vo, cloth..... 75

WARD (J. H.) Steam for the Million. A popular Treatise on Steam and its Application to the Useful Arts, especially to Navigation. By J. H. Ward, Commander U. S. Navy. New and revised edition. 1 vol. 8vo, cloth................................ 1 00

WEISBACH (Julius). Principles of the Mechanics of Machinery and Engineering. By Dr. Julius Weisbach, of Freiburg. Translated from the last German edition. Vol. I., 8vo, cloth.. .... 10 00

DIEDRICH. The Theory of Strains. a Compendium
for the calculation and construction of Bridges, Roofs,
and Cranes, with the application of Trigonometrical
Notes, containing the most comprehensive information in regard to the Resulting strains for a permanent
Load, as also for a combined (Permanent and
Rolling) Load. In two sections, adadted to the requirements of the present time. By John Diedrich,
C. E. Illustrated by numerous plates and diagrams.
8vo, cloth.................................... 5 00

WILLIAMSON (R. S.) On the use of the Barometer on
Surveys and Reconnoissances. Part I. Meteorology.
in its Connection with Hypsometry. Part II. Baro-
metric Hypsometry. By R. S. Wiliamson, Bvt.
Lieut.-Col. U. S. A., Major Corps of Engineers.
With Illustrative Tables and Engravings. Paper
No. 15, Professional Papers, Corps of Engineers.
1 vol. 4to, cloth................................ 15 00

POOK (S. M.) Method of Comparing the Lines and
Draughting Vessels Propelled by Sail or Steam.
Including a chapter on Laying off on the Mould-
Loft Floor. By Samuel M. Pook, Naval Construc-
tor. 1 vol. 8vo, with illustrations, cloth........... 5 00

ALEXANDER (J. H.) Universal Dictionary of
Weights and Measures, Ancient and Modern, re-
duced to the standards of the United States of America. By J. H. Alexander. New edition, enlarged.
1 vol. 8vo, cloth................................ 3 50

BROOKLYN WATER WORKS. Containing a Descriptive Account of the Construction of the Works,
and also Reports on the Brooklyn, Hartford, Belle-
ville and Cambridge Pumping Engines. With illustrations. 1 vol. folio, cloth.........................

RICHARDS' INDICATOR. A Treatise on the Richards Steam Engine Indicator, with an Appendix by
F. W. Bacon, M. E. 18mo, flexible, cloth......... 1 00

D. VAN NOSTRAND'S PUBLICATIONS.

POPE Modern Practice of the Electric Telegraph. A
Hand Book for Electricians and operators By Frank
L. Pope Eighth edition, revised and enlarged, and
fully illustrated. 8vo, cloth...................... $2.00

"There is no other work of this kind in the English language that contains in so small a compass so much p actical information in the application of galvanic electricity to telegraphy. It should be in the hands of every one interested in telegraphy, or the use of Batteries for other purposes.'

MORSE. Examination of the Telegraphic Apparatus
and the Processes in Telegraphy. By Samuel F.
Morse, LL.D., U. S. Commissioner Paris Universal
Exposition, 1867. Illustrated, 8vo, cloth.......... $2 00

SABINE. History and Progress of the Electric Telegraph, with descriptions of some of the apparatus.
By Robert Sabine, C. E Second edition, with ad-
ditions, 12mo, cloth...... .... 1 25

CULLEY. A Hand-Book of Practical Telegraphy. By
R. S. Culley, Engineer to the Electric and Interna-
tional Telegraph Company. Fourth edition, revised
and enlarged. Illustrated 8vo, cloth.............. 5 00

BENET. Electro-Ballistic Machines, and the Schultz
Chronoscope. By Lieut.-Col. S. V Benet, Captain
of Ordnance, U. S. Army. Illustrated, second edition, 4to, cloth................................. 3 00

MICHAELIS. The Le Boulenge Chronograph, with
three Lithograph folding plates of illustrations. By
Brevet Captain O. E. Michaelis, First Lieutenant
Ordnance Corps, U. S. Army, 4to, cloth.......... 3 00

ENGINEERING FACTS AND FIGURES An
annual Register of Progress in Mechanical Engineering and Construction for the years 1863, 64, 65, 66,
67, 68. Fully illustrated. 6 vols. 18mo, cloth, $2.50
per vol., each volume sold separately.............

HAMILTON. Useful Information for Railway Men.
Compiled by W. G. Hamilton, Engineer. Fifth edi-
tion, revised and enlarged, 562 pages Pocket form.
Morocco, gilt................................... 2 00

7

STUART. The Civil and Military Engineers of America. By Gen. C. B. Stuart. With 9 finely executed portraits of eminent engineers, and illustrated by engravings of some of the most important works constructed in America. 8vo, cloth.................. $5 00

STONEY. The Theory of Strains in Girders and similar structures, with observations on the application of Theory to Practice, and Tables of Strength and other properties of Materials. By Bindon B. Stoney, B. A. New and revised edition, enlarged, with numerous engravings on wood, by Oldham. Royal 8vo, 664 pages. Complete in one volume. 8vo, cloth....... 15 00

SHREVE. A Treatise on the Strength of Bridges and Roofs. Comprising the determination of Algebraic formulas for strains in Horizontal, Inclined or Rafter, Triangular, Bowstring, Lenticular and other Trusses, from fixed and moving loads, with practical applications and examples, for the use of Students and Engineers By Samuel H. Shreve, A. M., Civil Engineer. 87 wood cut illustrations. 8vo, cloth.............. 5 00

MERRILL. Iron Truss Bridges for Railroads. The method of calculating strains in Trusses, with a careful comparison of the most prominent Trusses, in reference to economy in combination, etc., etc. By Brevet. Col. William E. Merrill, U. S. A., Major Corps of Engineers, with nine lithographed plates of Illustrations. 4to, cloth...................... 5 00

WHIPPLE. An Elementary and Practical Treatise on Bridge Building. An enlarged and improved edition of the author's original work. By S. Whipple, C. E., inventor of the Whipple Bridges, &c. Illustrated 8vo, cloth........................ 4 00

THE KANSAS CITY BRIDGE. With an account of the Regimen of the Missouri River, and a description of the methods used for Founding in that River. By O Chanute, Chief Engineer, and George Morrison, Assistant Engineer. Illustrated with five lithographic views and twelve plates of plans. 4to, cloth, 6 00

MAC CORD. A Practical Treatise on the Slide Valve by Eccentrics, examining by methods the action of the Eccentric upon the Slide Valve, and explaining the Practical processes of laying out the movements, adapting the valve for its various duties in the steam engine. For the use of Engineers, Draughtsmen, Machinists, and Students of Valve Motions in general. By C. W. Mac Cord, A. M., Professor of Mechanical Drawing, Stevens' Institute of Technology, Hoboken, N. J. Illustrated by 8 full page copperplates. 4to, cloth.............................. $4 00

KIRKWOOD. Report on the Filtration of River Waters, for the supply of cities, as practised in Europe, made to the Board of Water Commissioners of the City of St. Louis. By James P. Kirkwood. Illustrated by 30 double plate engravings. 4to, cloth, 15 00

PLATTNER. Manual of Qualitative and Quantitative Analysis with the Blow Pipe. From the last German edition, revised and enlarged. By Prof. Th. Richter, of the Royal Saxon Mining Academy. Translated by Prof. H. B. Cornwall, Assistant in the Columbia School of Mines, New York assisted by John H. Caswell. Illustrated with 87 wood cuts, and one lithographic plate. Second edition, revised, 560 pages, 8vo, cloth................................ 7 50

PLYMPTON. The Blow Pipe. A system of Instruction in its practical use being a graduated course of analysis for the use of students, and all those engaged in the examination of metallic combinations Second edition, with an appendix and a copious index. By Prof. Geo W. Plympton, of the Polytechnic Institute, Brooklyn, N. Y. 12mo, cloth................ 2 00

PYNCHON. Introduction to Chemical Physics, designed for the use of Academies, Colleges and High Schools. Illustrated with numerous engravings, and containing copious experiments with directions for preparing them. By Thomas Ruggles Pynchon, M A., Professor of Chemistry and the Natural Sciences. Trinity College, Hartford New edition, revised and enlarged and illustrated by 269 illustrations on wood. Crown, 8vo, cloth.................... 3 00

ELIOT AND STORER. A compendious Manual of Qualitative Chemical Analysis. By Charles W. Eliot and Frank H. Storer. Revised with the Co-operation of the authors. By William R. Nichols, Professor of Chemistry in the Massachusetts Institute of Technology Illustrated, 12mo, cloth....... $1 50

RAMMELSBERG. Guide to a course of Quantitative Chemical Analysis. especially of Minerals and Furnace Products. Illustrated by Examples By C. F. Rammelsberg. Translated by J. Towler, M. D. 8vo, cloth.................................... 2 25

EGLESTON. Lectures on Descriptive Mineralogy, delivered at the School of Mines. Columbia College. By Professor T. Egleston Illustrated by 34 Lithographic Plates. 8vo, cloth....................... 4 50

MITCHELL. A Manual of Practical Assaying. By John Mitchell. Third edition. Edited by William Crookes, F. R. S. 8vo, cloth.................... 10 00

WATT'S Dictionary of Chemistry. New and Revised edition complete in 6 vols 8vo cloth, $62.00 Supplementary volume sold separately. Price, cloth... 9 00

RANDALL. Quartz Operators Hand-Book. By P. M. Randall. New edition, revised and enlarged, fully illustrated. 12mo, cloth......................... 2 00

SILVERSMITH. A Practical Hand-Book for Miners, Metallurgists, and Assayers, comprising the most recent improvements in the disintegration amalgamation, smelting, and parting of the precious ores, with a comprehensive Digest of the Mining Laws Greatly augmented, revised and corrected. By Julius Silversmith Fourth edition. Profusely illustrated. 12mo, cloth.. 3 00

THE USEFUL METALS AND THEIR ALLOYS, including Mining Ventilation, Mining Jurisprudence, and Metallurgic Chemistry employed in the conversion of Iron, Copper, Tin, Zinc, Antimony and Lead ores, with their applications to the Industrial Arts. By Scoffren, Truan, Clay, Oxland, Fairbairn, and others. Fifth edition, half calf.................... 3 75

JOYNSON. The Metals used in construction, Iron, Steel, Bessemer Metal, etc., etc. By F. H. Joynson. Illustrated, 12mo, cloth.......................... $0 75

VON COTTA. Treatise on Ore Deposits. By Bernhard Von Cotta, Professor of Geology in the Royal School of Mines, Freidberg, Saxony. Translated from the second German edition, by Frederick Prime, Jr., Mining Engineer, and revised by the author, with numerous illustrations. 8vo, cloth....... 4 00

URE. Dictionary of Arts, Manufactures and Mines By Andrew Ure, M.D. Sixth edition, edited by Robert Hunt, F. R. S., greatly enlarged and re-written. London, 1872. 3 vols. 8vo, cloth, $25.00. Half Russia................................ ..... 37 50

BELL. Chemical Phenomena of Iron Smelting. An experimental and practical examination of the circumstances which determine the capacity of the Blast Furnace, The Temperature of the air, and the proper condition of the Materials to be operated upon. By I. Lowthian Bell. 8vo, cloth.......... 6 00

ROGERS. The Geology of Pennsylvania. A Government survey, with a general view of the Geology of the United States, Essays on the Coal Formation and its Fossils, and a description of the Coal Fields of North America and Great Britain. By Henry Darwin Rogers, late State Geologist of Pennsylvania, Splendidly illustrated with Plates and Engravings in the text. 3 vols , 4to, cloth. with Portfolio of Maps. 30 00

BURGH. Modern Marine Engineering, applied to Paddle and Screw Propulsion. Consisting of 36 colored plates, 259 Practical Wood Cut Illustrations, and 403 pages of descriptive matter, the whole being an exposition of the present practice of James Watt & Co., J. & G Rennie, R. Napier & Sons, and other celebrated firms, by N. P. Burgh, Engineer, thick 4to, vol., cloth, $25.00; half mor........ 30 00

BARTOL. Treatise on the Marine Boilers of the United States. By B. H. Bartol. Illustrated, 8vo, cloth... 1 50

BOURNE. Treatise on the Steam Engine in its various applications to Mines, Mills, Steam Navigation, Railways, and Agriculture, with the theoretical investigations respecting the Motive Power of Heat, and the proper proportions of steam engines. Elaborate tables of the right dimensions of every part, and Practical Instructions for the manufacture and management of every species of Engine in actual use. By John Bourne, being the ninth edition of "A Treatise on the Steam Engine," by the "Artizan Club." Illustrated by 38 plates and 546 wood cuts. 4to, cloth...$15 od

STUART. The Naval Dry Docks of the United States. By Charles B. Stuart late Engineer-in-Chief of the U. S. Navy. Illustrated with 24 engravings on steel. Fourth edition, cloth................... 6 od

EADS. System of Naval Defences. By James B. Eads, C. E., with 10 illustrations, 4to, cloth........ 5 00

FOSTER. Submarine Blasting in Boston Harbor, Massachusetts. Removal of Tower and Corwin Rocks. By J. G. Foster, Lieut.-Col. of Engineers, U. S. Army. Illustrated with seven plates, 4to, cloth... 3 50

BARNES Submarine Warfare, offensive and defensive, including a discussion of the offensive Torpedo System, its effects upon Iron Clad Ship Systems and influence upon future naval wars. By Lieut.-Commander J. S. Barnes, U. S. N., with twenty lithographic plates and many wood cuts. 8vo, cloth..... 5 00

HOLLEY. A Treatise on Ordnance and Armor, embracing descriptions, discussions, and professional opinions concerning the materials, fabrication, requirements, capabilities, and endurance of European and American Guns, for Naval, Sea Coast, and Iron Clad Warfare, and their Rifling, Projectiles, and Breech-Loading; also, results of experiments against armor, from official records, with an appendix referring to Gun Cotton, Hooped Guns, etc., etc. By Alexander L. Holley, B. P., 948 pages, 493 engravings, and 147 Tables of Results, etc., 8vo, half roan. 10 00